= Mashurdrd chem =
Mas x the speed of
light x time

8
past the speed
of light +

$860.00 + 8 \times 2 =$
$860.00 + $ $2 =$

6050

13,000
3,050
358

$9 \times 0 + 1 = 820 + 1$

math

Get a toy. Get sandals.
Find a partner. Any
you
equation TV with a match
1000 0
105
with you

Never be afraid to love math.

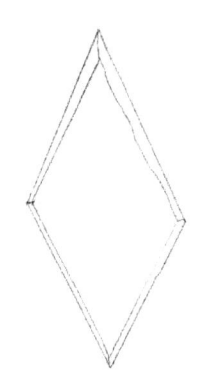

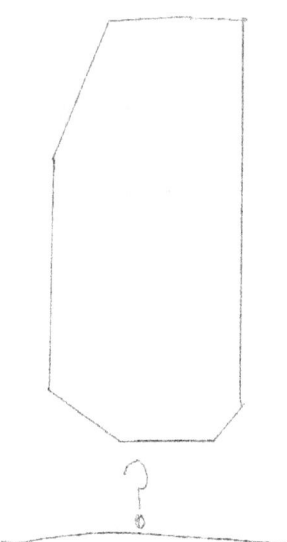

$$\left(E = M C \text{ squared} \right)$$

Evapor tlll

where am I

job llll

notes you take